Heavy Layers: The Methods Used by Tom Barron of England
or, *How To Get High Egg Production From Laying Hens*

by Tom Barron

with an introduction by Jackson Chambers

This work contains material that was originally published in 1914.

This publication is within the Public Domain.

This edition is reprinted for educational purposes and in accordance with all applicable Federal Laws.

Introduction Copyright 2017 by Jackson Chambers

Self Reliance Books

Get more historic titles on animal and stock breeding, gardening and old fashioned skills by visiting us at:

http://selfreliancebooks.blogspot.com/

Introduction

I am pleased to present yet another title on Poultry.

This volume is entitled "Poultry And Profit" and was published in 1921.

The work is in the Public Domain and is re-printed here in accordance with Federal Laws.

As with all reprinted books of this age that are intended to perfectly reproduce the original edition, considerable pains and effort had to be undertaken to correct fading and sometimes outright damage to existing proofs of this title. At times, this task is quite monumental, requiring an almost total "rebuilding" of some pages from digital proofs of multiple copies. Despite this, imperfections still sometimes exist in the final proof and may detract from the visual appearance of the text.

I hope you enjoy reading this book as much as I enjoyed making it available to readers again.

Jackson Chambers

BARRON-ESS II

282 eggs in one year, Second North American Egg-Laying Competition

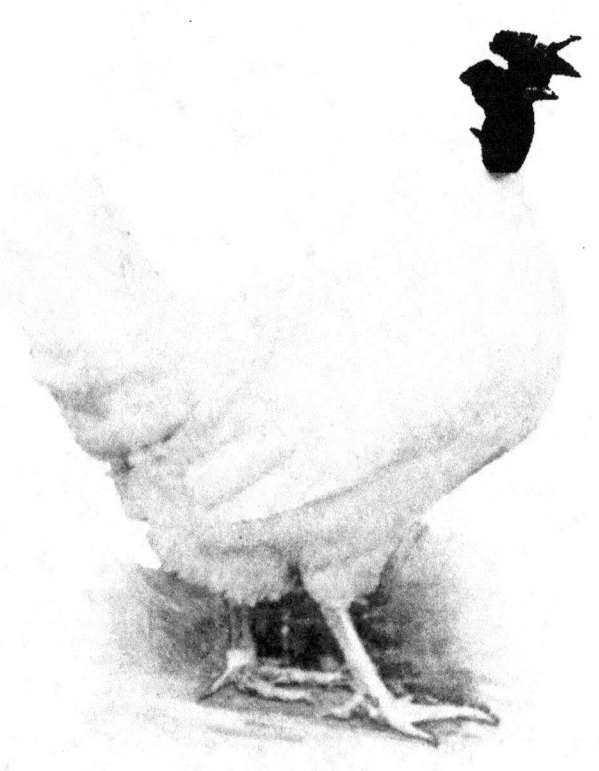

BARRON-ESS IV

256 eggs in one year, Second North American Egg-Laying Competition

INTRODUCTORY

Doing over nature, forcing the hen to lay eggs to an extent heretofore impossible has been the life work of the most justly celebrated of poultry breeders, Tom Barron, of Catforth, England; doing the very thing that nature made no provision for is the result of Tom Barron's work with hens, a work teeming with success to a degree that no American breeder has attained.

Manufacturing at will fowls of any selected breed or variety that lay more than the fabled "Goose of the golden egg," is to Barron as easy of accomplishment as is the failure of the ordinary poultry breeder to secure the two-hundred-egg hen.

England's poultry annals are emblazoned with the victories of this poultry wonder-worker; pullets that lay 282 eggs in 365 days, after shipment from their English home to America, five of them averaging 238 eggs in a year, under government supervision, between set dates, November 1st, 1912, to October 31st, 1913, and without recording any eggs laid outside of these inclusive dates.

The poultry world stands aghast at such records as have been made by the hens of this exceptional worker in egg production; since 1906 Tom Barron's layers have won prizes, almost numberless, in all of the English competitions and smashed every European, Canadian and American record for pens or individuals.

White Wyandottes, Buff Plymouth Rocks, White Leghorns and Buff Orpingtons, have all come under his magic touch. Think of six Wyandotte pullets, laying 98 eggs each, in six winter months! Again, a Wyandotte pullet is bred so that she lays 283 eggs from November 25th, 1909, to November 24th, 1910! Another in 1911-12 laid 263 eggs and at another competition still another laid 275 eggs in a recorded year!

Equally successful have been Barron's Buff Orpingtons, one splendid pullet laying 103 eggs in the 1910-11 Grimley, Worcester, England, 16 winter weeks' (112 days) contest, the four worst months of the year!

Buff Rocks have been just as pliable in Barron's hands, winning two competitions in England, his males used in breeding pens all being the sons of mothers with pullet records of 245 eggs each, in a year.

One of the Buff Rock pullets entered in The North American Third Competition, has a record of 125 eggs in twenty-three weeks. Needless to say it is direct from Barron's yards.

From 1906 to 1913 inclusive, in 11 different competitions in England and America enough silver and gold medals, silver cups and other valuable prizes have been won by Barron's birds to purchase a king's ransom.

Barron's winnings are given below in detail:

White Wyandottes

1907-8—First, Utility Poultry Club's Laying Competition.

1908-9—Second and First Class certificate and other prizes, Burnley Laying Competition, four pullets laying 256 eggs in sixteen winter weeks.

1909-10—Silver medal and Second Class certificate, Utility P. C. Laying Competition at Barron's Grange, Iden Rye.

1909-10—First Place, six pullets, 586 eggs, six winter months, Street Laying Competition, Somerset.

1909-10—Seventh prize, 574 eggs, six winter months, same event as above.

1912-13—Second Place, First Class certificate, four pullets, 249 eggs, four months, Northern Utility Poultry Society's Competition, Burnley.

1912-13—Third Prize, First Class certificate and gold medal at Harper-Adams Agricultural College, twelve months, Utility Poultry Club's Competition, at Newport, Salop.

White Leghorns

1906-7—Highest non-setters, Utility Poultry Club Laying Competition.

1908-9—Highest at Street (Somerset) Competition, averaging 93 eggs per bird, six winter months.

1909-10—Silver medal and Second Class certificate, Utility Poultry Club Laying Competition at Rye.

1909-10—Fourth Prize, best non-setters, Burnley Competition.

1909-10—Highest position at Street (Somerset).

1910-11—Second Class certificate, second highest non-setters, Grimley, Worcester.

1911-12—Second and Fifth Prizes, two First Class certificates at Burnley, 295 and 268 eggs, respectively.

1911-12—Fourth Prize, two silver medals, and other prizes in the Philadelphia North American International Egg Laying Competition. Two birds died during contest.

1912-13—Silver medal, First Class certificate, Utility Poultry Club, in the Harper-Adams twelve months Laying Competition.

1912-13—International First Prize, Second Philadelphia North American International Egg Laying Competition. Five birds laid 1170 eggs.

1912-13—International First Prize, cups and medals, Missouri State International Egg Laying Competition, Mountain Grove, Mo. Ten birds laid 2073 eggs.

Buff Orpingtons

1910-11—England's Champion Record, 103 eggs in 112 days, at Grimley, Worcester, in the U. P. C. Laying Competition, sixteen winter weeks.

Buff Rocks

1911-12—Fourth Prize, silver medal and First Class certificate, Utility Poultry Club Competition, Grimley.

1911-12—Second Class certificate was won by Buff Rock pullets at Burnley.

1913-14—First Prize, Third Philadelphia North American International Egg Laying Competition.

What Barron has done, however, is laid bare for the betterment of poultrydom.

His success was not dependent upon any superhuman knowledge; a clear objective, intense purpose, close observation and abject determination are responsible for his success.

Through generation after generation heredity found Tom Barron a shoemaker at his majority, his only asset a dream of better things. Persistent and provident, saving today that he might buy tomorrow, Barron grappled with fortune, good or bad, for better or worse, when his father, in 1904, turned over his little farm of three acres, two cows and twenty fowls, and Tom purchased it, retaining his cobblery against a possible rainy day. But it never rained, except the downpours were golden, and soon the farm grew from three to twenty-three acres, ten for the cows, which furnish the skim milk to rear the poultry, and thirteen for the poultry plant and home.

From the cobbler's bench to affluence through poultry culture is a far cry; Barron is a man in comfortable circumstances, his present easy condition having been brought about through his methods. The success of these methods warrant their adoption.

In this work his system of incubation, rearing and brooding, feeding at all ages, housing for breeding purposes, selecting the breeders, housing layers and feeding them are given in detail.

How Barron feeds his breeding stock is of the greatest value to beginners in poultry raising and can be put to the best of use by the experienced pultryman.

Beyond all other values or virtues attaching to this work the system of mating to produce layers in any breed or variety, stands monumentally before the public as the only purely scientific method of mating yet evolved for the purpose of egg production along ascending lines, whereby the greatest number of heavy layers will result. annually.

One year's work along the lines laid down by Barron will lay the foundation for an increased efficiency in any flock of fowls, good, bad or otherwise.

THE PUBLISHERS.

EXTERIOR OF BROODER HOUSE, SHOWING SMALL GRASS RUNS

THE
JOHN CRERAR
LIBRARY

CHAPTER I

Rearing Young Chickens

Close to my residence is a large brooder house—the chief one—which is 126 feet long and 16 feet wide. A picture of the inside and outside, with the runs, is shown here. A passageway down the center is four feet wide and doors of the little pens on either side open directly from this passage. With the exception of a foot of boarding skirting the floors, the whole length, forming the pens, is small mesh wire from roof to floor.

There are two distinct portions of the brooder house. One part containing the hovers is cut off completely by a wood partition from the other with a door through it in the passage. It is thus possible to keep the hover portion more free from draughts.

The chicks are brought from the incubator house in warm, covered boxes and put into this hover section.

The brooder system in use consists of hot-water pipe carrying boxes which run the whole length of this part of the passage. These boxes, eighteen inches deep, are cut off into sections opposite their respective little pens, which, in this portion of the building, are four feet square and accommodate some seventy chicks. These boxes are fitted to the floor. There is a little space between the bottom of them and the actual floor, allowing air to find its way in, keeping the floor of the hover cool and the air fresh.

The warm air having been breathed by the chicks leaves through the wire mesh at the top of the boxes so that a constant current of air is circulating through the hovers.

The floor of the hover is of wood, with sandy earth spread over it, and these floors although fitting fairly tight in the sections are movable up and down, nearer the pipes or away, as may be desired, by means of cog arrangements at the sides. These floors can be taken out, cleaned and the earth material renewed so that the floors are always sweet and clean, and being raised from the ground, have not the dampness of the earth.

The box sections are 2½ feet wide. One foot of this nearest the passage is wired over, and being hinged, forms an open topped lid so that the chicks can be easily put in or taken out. The other 1½ feet is over the pipes. Rods hold the felts which hang down in 1½ inch strips on either side of the length of the pipes. This also has a lid over it so that the felts can be taken out and cleaned or renewed frequently. The warm air is thrown down on the backs of the chicks. Being fed with dry feed at this time there is plenty of light through the net covered top to scratch for it.

The box sections open on the pen side, but of course until the chicks are to be allowed in the pens the small hinged door of the box falls back into the pen on an incline, but not steep, to the pen level.

Pen floors, which are cemented over, but not too finely finished, are covered with fine peat moss for scratching and are thus rough and deep enough to

INTERIOR BROODER HOUSE

keep the chicks from slipping and consequent sprains. Besides being well sky-lighted the room is also fitted with a window to each pen (of the hopper pattern) allowing for opening and the fresher air later on. As the chicks develop they are transferred to the somewhat larger pens on the opposite side of the passage.

It may be mentioned that for greater convenience the small pens attached to the hovers are only divided from each other by small partitions; this minimizes the ground draughts on the small chicks.

In transferring them to the other side of the passage they are still in the same temperature, but have more room which is equivalent to dropping the heat somewhat.

These and all the other pens in the building are the same size, namely, seven feet by five feet. Each pen has a box two feet each way, with a small door opening into the pen. Canvas or some thick material is dropped over the box and hangs down partly over the door sufficiently high to allow the chicks to run in and out freely. A small glass window in the passage allows one to see if the lamp burns properly, and, if needing adjustment, this can be attended to from the passage through a small hinged door in the wire mesh in the box.

In the middle of the box is placed the lamp with a steel circular guard enclosing it. A picture of the box, lamp and guard is given. The box is also covered with felt which conserves the heat as much as possible. This type of foster mother is adopted throughout the building and the chicks are transferred here as they

INCUBATING HOUSE

get on. In the outer wall of each pen there is a small door sliding in grooves and lifted by rope and pulley from the passage. This gives access to the grass runs. These runs are seven feet by fifteen yards long, one for each pen. These are turned over each year and sown down with grass seed so that a fresh growth is ready for the spring hatched chicks. The chicks have always done well on this system.

The young chicks are fed on grain until they are a fortnight old, the dry feed being thrown into the peat moss, chaff or other litter. The soft feed then given is fed to them on stiff cardboard, or pieces of wood. This saves waste. The pens are cleaned out, well limed, and renewed with litter each week. Cleaning and liming is absolutely necessary. By doing this there is no such thing as gapes.

There are eighteen pens on each side of the house and each of these is lit by acetylene gas from a jet about two feet from the pen level. The passage is also fitted with jets at regular intervals. In the dark winter months, when the chicks have been in the building a few days, it is lit up from about seven in the evening until midnight, when the gas is automatically turned off. This means that the house having a good light for several hours longer the chicks take so much more exercise scratching for their feed. We have found that this reduces leg weakness to a minimum. The chicks quickly take to artificial lighting and are soon out and working.

In each of the larger sections, that is, other than the hover pens, perches are put in so that the birds are induced to perch early. I believe in this method.

The drinking pots are raised slightly on pieces of wood which keeps out the scratching litter and the water fresh. A great point is to frequently change the drinking water.

It is advisable to keep the size of the chicks in a pen about the same all round. If there are one or two which have not come on quite so quickly as they might, but are still healthy, they should be put back with chicks somewhat younger and more of their present size. Small chicks with large ones never get a chance. They are constantly trampled upon and pushed aside from the feed.

COLONY HOUSE FOR REARING YOUNG CHICKENS, WITH BROODER BOX USED IN IT; ALSO LAMP, GUARD AND SHALLOW DRINKING POT

CHAPTER II

Another System of Rearing Chickens

We have on this farm about fifty chicken-rearing colony houses with some 100 grass runs attached. A picture of one of these is given with the box, lamp and steel guard used in it. The houses are seven feet long by four feet wide. They are five feet high in front and four feet three inches at back. The roof in front overhangs some six inches and weather porches stand out over the side windows to keep out driving rain or snow. It will be seen that these are set out on the passage plan similar to the breeding pens. Each house has a run to itself of ten yards by fifteen yards, in which we put, say, about sixty or seventy chickens.

We put a box which fits in one end of the cabin in place, the front of the box facing into the cabin being for the most part open. In this the lamp is placed with a circular guard enclosing it and over the box is dropped a canvas or some thick material and let fall over the opening, but sufficiently high to let the chickens run in and out to the warmth. At each side of the door in front of the house, always facing the passage, are sliding windows, the openings being covered with wire mesh. When the weather is suitable and the chicks are old enough these windows are dropped, giving them as much fresh air as possible. The lamps are kept burning all night on the cold, dark nights.

When the chicks are large enough to do without the heat it is withdrawn, the boxes removed and perches put in, the rests for these being already in position when the house is built. Peat moss and chaff is spread over the floor, the chickens being fed with dry feed thrown into this.

When the birds are old enough and the hovers are removed, wirework frames are also put into the cabins, the bands of the frames, cutting off the angles of the buildings, keeps the chicks from the corners at night.

Each grass run is divided from the next one by boarding sixteen inches high and netting above this a further three feet. The boards check the cold winds considerably. When the cockerels are maturing this barrier stops them from damaging one another, as they will do when separated by wire mesh alone. Where the grass is uneven in the run near the cabin door, turf is put down to level it up somewhat, otherwise the young and tender birds are apt to sprain their limbs.

In very hot weather any drinking vessels anywhere on the farm, and particularly amongst the young stock, if in any way foul are entirely emptied and filled with fresh water. I believe any amount of illness in birds during summer could be traced to carelessness in the matter of fresh water. There is no doubt that where the birds are watered by streams and these have run dry the birds are frequently poisoned.

PARTIAL VIEW OF FIFTY REARING PENS. LARGE SEMI-INTENSIVE HOUSE IN BACKGROUND

CHAPTER III

Feeding Young Chickens

After hatching in the incubators on this farm the chicks are not given any feed whatever until they are twenty-four hours old. From that time until they are a fortnight old they are fed on dry feed which is composed of the following:

> Wheat, four parts;
> Canary seed, one part;
> Red millet, one part;
> White corn grits, one part;
> Red corn grits, one part;
> Rice, half part;
> White millet, one part;
> Hemp seed, quarter part;
> Ants' eggs, half part;
> Dried flies, half part;
> Ground kibbled marrow fat peas, half part;
> Pinhead oatmeal, two parts;
> Grit, five per cent.

This mixture is thrown into the litter after they have learned to pick by putting the grain on a stiff cardboard. The scratching out of the dry feed I consider most essential to the health of the chicks.

Plenty of grit is kept before the young birds and the water is constantly changed for fresh.

ILLUSTRATING WELL-GRASSED YARDS

When the chicks are a fortnight old this feed is dropped in favor of soft feed, prepared in the following manner:

On my farm are ten milch cows, and the milk from them is run through the separator. After the cream is taken away, the skim milk is put in reserve for the chickens. This fresh sweet skim milk is put into a forty-gallon boiler and into this is put a large bucketful of rice.

Some ten to twenty gallons of this milk will be boiled at a time according to requirements. It is steamed until the rice and milk look like one large milk pudding. A galvanized tin, in which the stuff is boiled, is carried to the mixing trough and the contents poured on the top of about thirty pounds of biscuit meal. This dust is useful in that it not only is in itself a good food, but it keeps the rice and milk from forming into a stiff mass, splitting up the grains, as it is mixed in, into small crumbly pieces. When run through a sieve these are about the size of a pea, suitable for the chicks to pick up without waste.

The mixture must on no account be sloppy. If found too wet a little more of the biscuit dust will dry it off.

I do not believe in giving young chicks a wet sloppy mixture. If anything keep on the dry side.

This is given each day at noon.

After the chickens are a month old we feed half rations of the chick feed, given above, and mix in another half of split Indian corn, a few ordinary groats, and wheat, and make it a cheaper food eventually, gradually getting it to groats, ground oats, wheat, buckwheat, Kaffir corn, and as many mixtures as possible.

At the age of three months the chickens are divided into their respective sexes, the males being placed elsewhere.

INTERIOR OF SEVENTY-TWO-FOOT LAYING HOUSE

THE
JOHN CRERAR
LIBRARY

CHAPTER IV

Housing Layers

Our layers are housed in what we call, in England, the semi-intensive system. The flocks are distributed, some being housed in small buildings similar to the ones used for breeding purposes, and others in very large houses. A picture of one of the larger buildings showing the inside and the outside, with the birds, is given elsewhere. The dimensions are seventy-two feet long, eighteen feet wide, eight feet to eaves, and thirteen feet to gables.

In one of these 400 Single Comb White Leghorn pullets are accommodated and a flock of White Wyandottes is housed in another similar house. It has a span roof, boarded underneath, with corrugated iron-work covering the woodwork.

In the south wall are fitted shutters practically the whole length, which, by means of ropes and pulleys and counter-balance weights, can be lifted or let down at will, according to the weather, the openings being covered with wire mesh. This makes the south side almost open fronted.

A storm screen, three feet wide, fixed to the roof outside is immediately above these openings which protects the floor of the house from driving rain, or from the full power of the sun's rays in close or hot weather, as the case may be, but allows of plenty of fresh air circulating through the interior.

The nest boxes, of which there are thirty-eight, are fourteen inches each way, and are situated just below the windows.

The tops slope sharply from the wall to front of nest, keeping the birds from using this part as a perch. On the opposite side of the house is a dropping board running the length of the building, five feet broad, and two feet six inches from the floor.

The perches are themselves three inches deep and one foot above the dropping board. They are arranged transversely along this, being eighteen inches from center to center.

Long perches incline the birds to crowd together on cold nights, to get overheated, and when they move, to catch cold. Being split up into forty perches across the board distributes the birds more.

Every twelve feet this dropping board with its perches above is divided up from the floor to the square of the house. The whole length above is lathed over and covered with straw and this stops side and down draughts winter nights. Partitions can be taken out for the warmer months.

At each end of the building is a door eight feet wide, running on overhead pulleys, allowing a horse and cart to pass through from end to end for cleaning and other purposes.

The house generally, and the floor in particular, is well-lighted from the roof with twelve windows let into the wood and iron, three feet by two feet, also along the bottom of the north wall, under the dropping

SEMI-INTENSIVE HOUSE, SEVENTY-TWO FEET LONG, HOUSING 400 LAYERS

board by some fourteen windows, twenty inches square.

The birds have, therefore, no difficulty in scratching out their grain feed from the litter in which the dry stuff is always thrown. The litter is always three to four inches deep, of cut straw, peat moss, and chaff.

Water is contained in two large wooden piggins mounted on platforms to keep out the scratching litter. These are emptied entirely, scrubbed and refilled each day, and more frequently during hot weather, if fouled.

Mash feed towards evening is given the birds in the troughs directly on the floor, but the green feed of mangles, turnips and cabbage is put into specially made troughs, mounted on trestles, with a standing board for the birds on each side of its length. Sometimes the green food is strung up within reach of the birds. This gives them fine exercise, particularly during the winter months.

In addition to the feeds given, dry in the morning and soft mash at night, the birds always have before them a bran hopper which contains one part ground oats and three parts bran. Grit and shell are also constantly available.

The floor of the house is raised about six inches above the ground level outside so that no matter how wet the weather the floor inside is dry. The floor of these very large buildings is made as follows:

About twenty tons of cinders is put down, rammed, rolled hard and leveled up into as perfect a surface as if the floor was finished. Then four tons of asphalt are laid over it and hardened.

When this is ready a layer of fine cement gives a finish, without being fine enough to cause sprains to the birds' limbs. It is laid about one inch thick. This forms one of the best floors I have ever used.

Runs are on both sides of the house and are reached through the end floor windows which slide back, two similar ones having been also fitted in the south wall. These runs are used alternately. When the birds have been using one side for some little time they are left out in the opposite grass run and the other allowed to freshen. The two sides form a one-acre grass plot. My cattle are turned into these pens alternately and help to keep the grass from getting too long.

ONE OF THE LARGER LAYING HOUSES

CHAPTER V

Feeding the Layers and Breeding Stock

My system of feeding the layers and breeding stock is much the same as regards the feed given all the year round, any difference made being chiefly in the amount fed according to season and my own discretion.

In the morning we feed all the birds by hand with a mixture composed of

    Wheat,

    Split Indian corn,

    Coarse groats,

    Kaffir corn,

    And other grains to mix with these according to what can be bought reasonably cheap in the market at certain times.

But the more kinds the better, as the birds like changes. It keeps them working better and while they work they keep from a multitude of ills.

This is given fairly early in the morning according to the light, thrown into the litter and sufficient is given to keep the birds working most of the day. If not working, about midday a little more grain is given, not a meal, but just something to keep them going. We feed fairly heavy with this feed, not stinting them, but allowance where possible is made for the weather, for if fine enough, they are better out foraging for the grubs and bugs. But the point is that the dry feed does not lay heavy with them and cause them to be sluggish. Besides it keeps them constantly occupied scratching the feed out of the litter.

At night we give a wet mash. Of course, many breeders do not believe in this system, but reverse the order, giving the warm, wet mash in the morning, the argument being that the birds require something easy to digest quickly as they may not have had a feed since the afternoon previous. My plan is based on years of experience and I find that with a morning fed mash, especially on cold days when other work is necessary they get lazy, hop out and stand about on one leg half the day, getting chilled thoroughly. The grain feed has a tendency to produce the opposite effect and the warm mash late in the day makes them feel comfortable and restful for the night.

We do not feed the mash very wet, but on the crumbly side. It is made as follows:

> One part bran,
> One part biscuit dust,
> One part middlings,
> Half part ground oats,
> One eighth part clover meal,
> One eighth part fish meal, or meat meal.

With this they are fed the year round. This is given to the birds in their troughs and as much as they will eat. If we consider the birds are getting too fat the difference made is that we scald the bran. If on the thin side, then we add a portion of corn meal. The scalding of the bran will open the bowels of the birds and so reduce flesh without injuring their general condition.

A MORE DETAILED VIEW OF REARING PENS

The mash is prepared as follows:

A boiler holding forty gallons is filled with water and into this is put the clover meal and about three bucketfuls of oats or wheat. This is steamed well, but not necessarily boiled. The bran, biscuit dust, ground oats and Pollard, are put into the mixing trough and mixed well. Then the steamed mixture from the boiler is turned over this and the whole lot mixed together. The mass is then mixed dry or crumbly with middlings and meat. After this, if the mixture is still too wet, we add some more biscuit dust, until the mash is crumbly, not sloppy.

In putting the clover meal, etc., from the boiler on to the bran, biscuit dust, etc., in the trough, you will find that the heat is not sufficient to scald the batter thoroughly. More of the nutriment of the feed is therefore left in it than if properly boiled. This keeps the birds in good condition. Should the birds get too fat then scald the bran as a wet mash by itself. You will find that this will purge the birds and so reduce the fat.

To the mixture in the trough mentioned above about twice we had a roup powder consisting of one part epsom salts, one part magnesia, one part sulphur and one part ferri sulph. and gave in the proportion of a gill to every 100 birds. This prevents a good many colds the birds would otherwise have, and is a splendid cure for roup.

If there has been an outbreak and a severe one, of colds, add to each gallon of the drinking water as much sulphate of copper as can be put on a twenty-five-cent piece. This powder, which is really a tonic, helps the birds to throw off the effects of the disease. It is also good for young chickens. About half the amount only, once or twice a week should be given the youngsters. I consider this roup powder to be the best obtainable.

The birds are fed all the year around with green stuff although more is given them during the colder months when they are unable to get as much for themselves. A mangle is given usually each day to every twelve birds.

There is nothing worse than sun-warmed water, and water that has gone slightly green through being stale. This green deposit which settles on the sides and bottom of the vessel is scrubbed off regularly. We do not mind if a little of the disinfectant, with which the vessels are cleaned, is left in the pot afterwards, as it does the birds no harm, but good.

INTERIOR, BREEDING HOUSE

CHAPTER VI

Housing for Breeding Purposes

A picture of the inside of one of my breeding houses, and exterior passages, is shown herewith. Each house is twelve feet long, nine feet wide, five and one-half feet to eaves, and eight feet to gable.

It will be seen it has a rather steep roof which runs the rain off quickly and preserves the material a great deal. If the roof is battened, or stripped over the joints, it does not require felting. Personally, I have a dread of felting as after each storm there is such a job repairing damage.

The house is fittted with a dropping board, three feet from the floor and one foot above this are the perches. The dropping board is 2¾ feet wide, and has a ledge on the outside edge to keep all material, such as peat moss and chaff, from falling to the floor.

Taking the perches back to front, the first one is one foot to center, and the next one fifteen inches to center. This does not overcrowd the birds at night. The perches being on the same level stops the birds from fighting as they do when one perch is higher than the other.

A partition of laths divides the house in two and a door allows passage through the house to trap nests, etc. The laths in the partition are closer together near the ground to obviate any accident to the birds by getting their heads fast.

In one line on the south side are fitted the trap nests, 2¾ feet from the floor. The dropping board and the nests being well above the floor checks the birds from laying on the floor, as they do with shelves too near the ground.

Below the board and the nests, that is in the north and south walls, are four windows, eighteen inches by thirty inches. There are also windows let into the gable ends. These thoroughly light up the floor, giving the birds every chance to scratch out the dry feed thrown into the litter. This light is necessary as the house is a scratching shed and perching shed combined.

On the south side is an open front three feet long by eighteen inches deep, covered with wire mesh (let down or pulled up, as the weather may require, by rope and pulley from the passage between the houses). A small door opens into the run slightly above the ground level, through which the birds reach the grass pens. The door slides in grooves and is worked by rope and pulley from the passage. The birds are thus let out without entering the runs.

Each house has a grass run of 20 yards by 24 yards and two runs adjoining immediately opposite the middle of the building gives each pen of birds a run of 10 yards by 24.

In the early spring ten hens and one cockerel, White Leghorns, are put into a run; (eight females and one male, if White Wyandottes). We have found this system of mating, and size of run and house, do well for fertility. This last season seventy-five per cent of all eggs put into the incubators hatched out, the eggs not being tested beforehand. Eighty-five to ninety per cent were fertile.

Another system we tried this past breeding season was the double mating which did very well. The arrangement is that, say, fifteen White Leghorn pullets are mated with two brothers, but one cockerel is made to rest for five days in a special spare cockerel coop (see picture), while the other male runs with the birds.

The reason for this is that we have found that one male will starve itself, even to death, to feed its females, but the enforced rest brings them back into condition again. They are also more fertile on this plan which is perhaps only natural.

Now, as to planning the cabins, take one of the breeding house lots as a type of the others. It has twelve houses on each side of the passage with their twenty-four grass runs. The houses have their length parallel with the passage, but the length of the runs is at right angles to this, saving a lot of walking. The alley is $13\frac{1}{2}$ feet wide, and although the houses, on one side at any rate, stand in this, there is still enough room to allow of the easy running of a horse and cart through the whole length.

Doors of the houses open into the passage so that it is possible to walk almost straight along through one side of the series, thus facilitating feeding, trap nesting and other work. The birds are also watered directly from the alley, the troughs standing in the passage. The birds reach the water through a wooden frame fifteen inches square, with round upright bars, $2\frac{1}{2}$ inches apart. The frame is fixed upright to the ground in line with the netting of the pen. Hence, birds do not crowd each other out when drinking and it checks them standing on the edge of the pot and fouling the water. An illustration shows this excellent arrangement.

The breeding coops on legs one foot long and with floors of round bars fairly wide apart and roofs slightly sloping from front to back are virtually all in line with the cabins set in the passage, and handy for the removal of would-be sitters from the breeding cabins. This places the birds in a position to see what is going on and helps to break the broodiness.

Fruit trees are planted in the runs and these afford shelter to the birds from the sun's rays. The grass runs are boarded on each side, two feet high, wire netting three feet on top forming a division five feet high. The boarding protects the birds from strong winds. It also stops the male birds from fighting, as they will if they can see and get at each other through wire mesh. Fighting brings down their condition and the result is more infertile eggs.

As regards ventilation in the houses, it might be mentioned that the windows in the gables are hopper pattern and when the weather is suitable are thrown open. There is also an inch special ventilation space between the actual roof and the ridge capping.

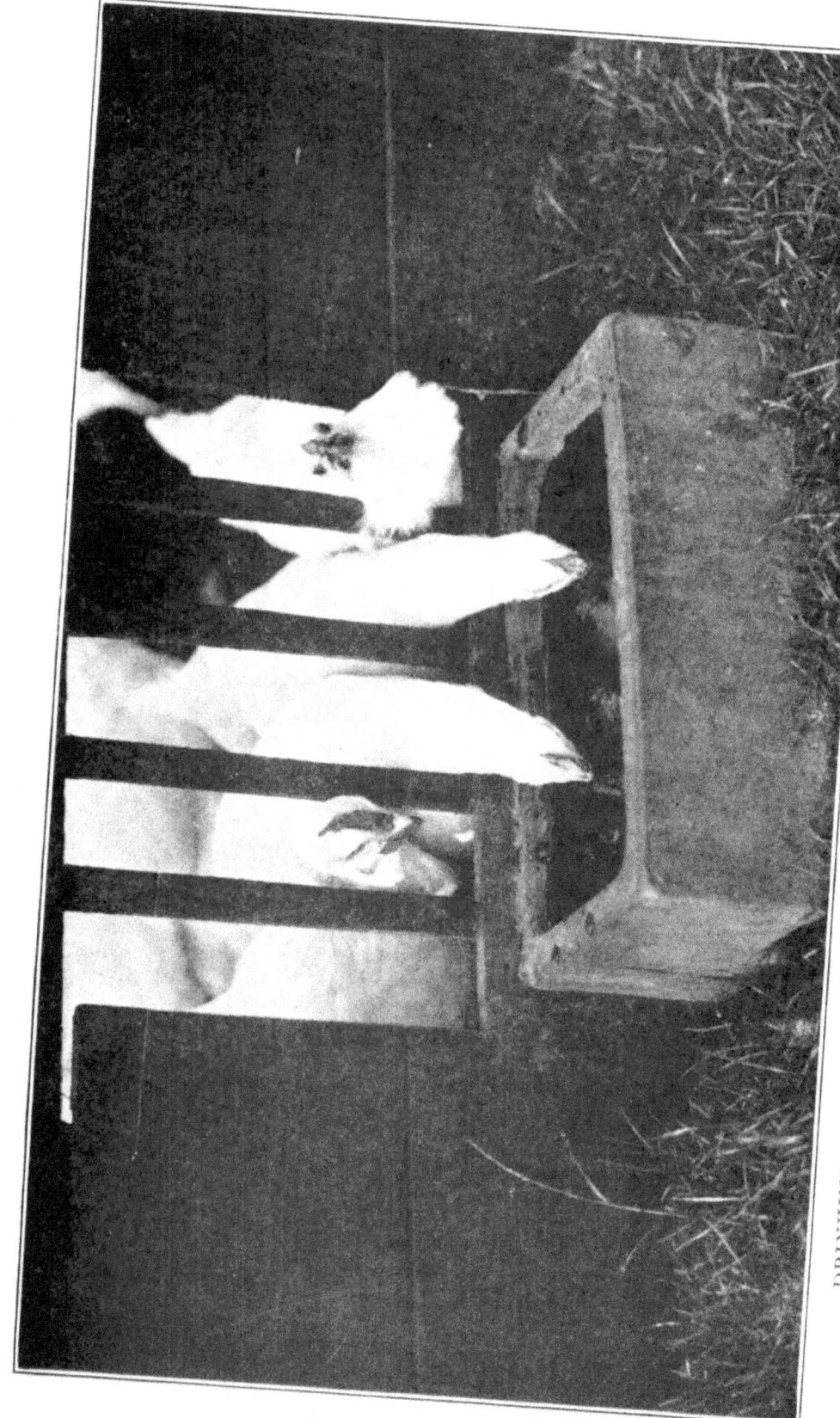

DRINKING FOUNTAIN IN PASSAGE WAY. NO CROWDING WITH THIS ARRANGEMENT

THE
JOHN CRERAR
LIBRARY

CHAPTER VII

Selection of Stock

As a young fellow in the country I had seen any amount of poultry kept on the general farms. The prevalent idea was, if the hens laid, well and good, but if not, it could not be helped. But this was not my way. Being passionately fond of poultry keeping I put my whole mind and soul into it.

First, a few mongrel birds were kept and account taken of the sales. They paid. I got a little money together by care and bought some stock. But here my first real lesson in the poultry business came in. I purchased some cockerels to mate with my best layers. They were gotten from a noted English exhibition breeder who produces both layers and exhibition birds. These were mated with the layers that I had picked with my own eyes.

The progeny were the worst layers I ever had. It would have been far better had I given the breeder $1000, than to have bought his birds. I decided to use the trap nest and did so from then on.

But the point is that fine feathers do not make good layers. If, as I believe is frequently the case, the birds are culled out by the breeder from amongst his prize birds and sold as utility birds, for breeding layers, they are worse than useless. It is my advice to anyone to get the best they can to start with, even if they have to get less in number.

As to the best breeds I do not want to tread on anyone's toes, and I know I am on critical ground, but in my opinion the White Leghorns, White Wyandottes, Buff Orpingtons and Buff Rocks, are the most profitable utility birds.

If you cross White Leghorns with White Wyandottes, using a cockerel of the former by preference, then you have one of the best crosses for eggs that I know of.

Of course my business demands that I shall have these breeds as pure breeds on my farm, but I write from previous experience and also know of one among many instances where a farmer who had 125 of this cross in one pen for twelve months, got eggs from them which averaged to the value of $5.00 per bird, for the year.

As to knowing the best birds for laying by appearance alone is chiefly a matter of personal judgment, only obtained after years of experience. But some of the best may be selected by observing the pictures of the noted layers shown in this book. I may say that I do not think it possible to get both perfect layers and birds for table, or for show, whatever breeds are used, the best of one is produced only at the expense of other qualities.

Egg laying is a drain on the system which produces muscle in the bird, but does not produce flesh suitable for the table. Show points of a breed do not go hand-in-hand at all with the best type bird for laying.

SINGLE-COMB WHITE LEGHORNS, SELECTED AS THE TYPE TO PRODUCE LAYERS

However, I breed entirely for layers and birds to breed layers, and take this view completely in this book. The only safe way for the beginner is to procure stock from a breeder who has systematically trap-nested and selected his birds for generations.

If I were selecting a pen of layers for a laying competition, first of all I should certainly know how the birds had been bred for several generations. They would have to be bred from hens that had laid over 200 eggs in their pullet year; also they would need to be sired by the son of a hen that had laid over 200 eggs.

But independently of this when you get a pen of sisters, bred from one sire and one dam, you will find they will vary a great deal in appearance.

In selecting from these I should pick out the ones with the longest back, the broadest back, the longest breast bone, rather of a wedge shape, with the greater part of body behind her legs; a sharp eye, fine texture of comb, and rather straight narrow skull. I like the eye to be almost protruding from the socket, as if the birds were staring. And the tail would be inclined to go forward towards the head. Some people call this squirrel-tailed. Birds suit me best short of leg.

I believe in the small bird, as I have found after long experience that the large bird does not lay the larger egg, nor as many of them, as the small bird.

In the selection of stock for breeding purposes, with the object the production of eggs, pure and simple, there are laws of heredity to be considered which are quite as decided and strong in effect, probably, as in "fancy" breeding, and these must be observed generation after generation to get the tip top layers.

It will be found that certain of the pullets do not lay as well as their sisters do, and their dams did, but yet lay eggs of a larger size, better shaped, and color. Then the idea is to pick out the birds that lay the most and best eggs, and mate up with the cockerel showing the best cockerel utility points.

It is often said that "like breeds like," but there are modifications. Some of the cockerels may have bred too long in the leg; this must be considered in the next mating.

My idea of the best male for producing the best breeders of layers in the White Leghorn is a sprightly bird, not too large, with good straight comb, active, long and broad breast bone, not too high in leg and the longest part of the body behind the legs. But these outward signs must never overshadow his recorded pedigree. His toe marks when examined must show that his parentage on both sides is from birds with not less than 200-egg record for the year. No cockerel heads a breeding pen of mine unless his record indicates this.

As to color, I do not worry much about that in selecting for breeding pens. He may be a bit brassy, or sappy on top, but if he is a good breeding bird I do not allow this discoloration to weigh in my decision.

Unfortunately, some people, and I say not a word against them, for they have an equal right to their opinion, although wanting eggs badly, still cannot get the idea of "fancy" out of their heads. I have seen many instances where people have passed over a really good breeder, because the bird was not a pure white, and select a much inferior bird, from the utility point of view.

The question of birds being nicely shaped and good colored, does not influence me at all in taking up certain breeds. I have on the farm at the moment of writing, four breeds, but only after I had proved to my own satisfaction that White Leghorns, White Wyandottes, Buff Orpingtons and Buff Rocks would suit me as the best laying breeds.

Many people have visited my farm, both Britishers and Americans, representing the "fancy" as well as utility poultry. They have said that my birds are good looking, good shape and color, and a great many of them typical (and this is backed up by customers who have favored us with testimonials) but I say again, they are not bred for show purposes, only for laying, and sold as such.

I do not show myself, for if I did I should never get a prize. The reason for this is that our exhibitors in England have bred for the bird to look at, never considering that a bird had to lay. Size is the greatest point with the English exhibitor. Size, color and other silly things are carried too far.

The trap nest tells us that the smallest birds are the best layers, and the best cockerels to breed from

WHITE WYANDOTTES, BRED FOR MATING TO HEAVY LAYERS

are not the largest. Then why in the world, in the exhibition birds, should they choose the largest birds to win?

Take the White Wyandotte. The exhibition bred bird is short, cobby, shortbacked and of a tremendous size. Now the same bird for utility, the Wyandotte that the utility men of England use and breed especially for, is the small bird, long in the back and short on leg. These are good enough to lay over 200 eggs per year, while the exhibition specimen will probably not lay more than 100 in the same time.

See what this means in a large number of birds?

CHAPTER VIII

Trap Nested Birds

Whether everyone agrees with trap nesting layers or not, it is quite certain that breeders who adopt utility breeding, no matter what breed or system of mating they favor, desire to know which pullets are the best layers, and which cockerels have the best pedigrees. It will be apparent to the experienced poultryman that the most careful observation of a large flock fails to show exactly which are the best layers. It is impossible to keep count of all. It follows then that a recording system must be adopted to pick out the best layers and determine their year's record.

Some twelve years' practical experience in the use of the trap nest has convinced me that this system has yet to be improved. Note also in how many other ways it is useful and pays for the time expended. Yet, very few breeders, a handful at most, have gone in for it. It is the sum total of economy to the small man, but what to the large? The latter cannot be calculated. Mammoth firms cannot afford to provide tons of provender for birds that eat, but do not lay.

Say a man has bought some birds. He may look them over critically, but if they do not come to him with their records, or pedigrees, he has no idea of their capabilities. But supposing we give him credit for a certain amount of judgment and the birds he has bought do actually lay an average of 100 eggs in the year. Why should he not have paid a dollar or so more and doubled his egg output, and that not just for the first year? Look at the result of purchasing some good trap nested layers, on a farm carrying, say 5000 layers for market eggs only, and in the United States there are a great many topping this.

Consider the satisfaction of buying trap nested birds, and pedigreed cockerels, for the day-old chick business, when the buyer knows that the parent birds had a definite pedigree for fertility (the male being bred from vigorous stock for generations), and the male's dam and his sire's dam had actually laid over 200 eggs each, in the year.

The man also, purchasing eggs for custom hatching knows by this means that he will have a reliable return for his outlay.

Some people have the mistaken idea that trap nesting the birds upsets their laying capabilities, and that not so many eggs are laid. It does not work out so in practice, but rather the reverse, from the extra care and watchfulness they actually receive. The laying of a particular pen of birds is virtually a guide as to the amount of feed to give them, the season and other points considered. Should a bird unseasonably stop laying for some time, it would point to some defect or illness in the bird, and the breeder can then get to the bottom of it. But how could he put his finger on the waster without such record? The bird laying the undersized, oversized, misshapen or bad colored eggs is soon identified.

An experienced man in handling the birds frequently has good opportunities of noting these troublesome disorders which tend to reduce the eggs laid without stopping the output entirely. It is also an index as to the pullets that have come on quickest to lay, and is a guide for the future in picking out these birds which will lay just at the time eggs are earning the biggest prices.

An item, which on a big farm may become a large saving or leakage is the broody hen. By using the recording sheet to show when the hen was taken off, and when put back to the nest, birds will not be kept off unduly and some eggs lost.

But to the man who is line breeding seriously for utility purposes, this trap nest system, it seems to me, is indispensable. I cannot see how he would do without it. To carry the record of about ten generations of a score of birds in his head would mean a remarkable memory indeed.

I have used two or three types of trap nest shown here, but the one in general use at the moment, although some may yet improve upon it, is a good one. Each nest box is twenty inches from back to front, and twelve inches side to side. The shutter at front slides down in grooves and is four inches deep, with legs to keep it from sliding fully to the bottom.

When the nest is open this shutter is held up from the inside by a frame, with its pivots resting in sockets inside the box, at the sides.

The top part of the frame is weighted and catches on a small ledge on the inside of the shutter. The bottom part of the frame, which just works clear of the actual nest section into half turn, forms a treading bar. The hen in entering the nest steps on the broad treading bar. Her weight pulls the bottom of the frame towards her, bringing away the top part from the shutter, which is then released and falls gently behind her. The frame then turns back again. A piece of wood 1¾ inches high divides the box floor in half, the back portion for the nest proper and front space for the swing of the frame. The small strip also keeps the straw of the nest in place.

The hinged lid with just a bit of fixture at back virtually covers the top of the box with half inch over the front. This lid, when opened, falls back to the wall; when closed (in my newest type), slopes sharply from the wall to the front. This stops the birds using it as a perch at night.

The trapped bird is easily taken out of the top of the nest and its band number checked on its record sheet. It will be seen that the bird is actually well into the box when she treads on the bar and the shutter falls, so there is little noise to frighten her, and no danger of hurting her tail when the shutter falls. The division, the frame and shutter being all removable, leaving a plain box, can be readily cleaned, sprayed, and has no crevices. The nests are so conspicuous that the pullets use them at once and there is no loss in individual records.

Chopped straw about a foot long, fresh, clean and dry, is used for the nest itself. This is changed frequently in the summer when poultry enemies are most numerous.

My recording cards each contain a full month's record in days and total. There are some fifteen sheets attached to the stiff card which can be turned over round the bottom of the card where they are stapled to the back, so that each bird's full year's record may be taken on one card. The month's record is carried forward on the next sheet. When the card is completed the totals are taken into the recording ledger in the office where the points of percentage, the mating of the pen, unmarketable eggs, moulting periods and misadventures are entered.

If a man has a stock of 5000 layers and has never used a cockerel known to be bred from a hen that has laid over 200 eggs in the year, it is quite time for him to begin to do so. If he would only trap nest a few pens and breed from the highest laying hen, provided, of course, she has health, vigor, size of eggs, color of egg and stamina on her side, he would marvel at the results. Cockerels bred from such a hen, mated to large flocks, will produce from twenty to fifty eggs more per bird per year than if the selection of the layer is done in the old-fashioned way.

I think this is a great point, and too much stress cannot be laid on the matter. On my farm we trap nest a dozen pens of each breed, and these are mixed differently each year, so that unrelated stock, as near as possible, is produced each season.

THE
JOHN CRERAR
LIBRARY

CHAPTER IX

Mating to Breed Best Layers

The way that my birds have been laying that have won the great contests—the Philadelphia North American International Egg Laying Competitions and National Egg-Laying Contest at Mountain Grove, Mo.—have naturally aroused a lot of interest. The question has arisen in a hundred and one directions, "How is it done?"

I will tell my readers.

As an example, I will explain the way I bred the birds in these competitions. For illustration and to make it less complex, I will take just one or two pens; the principle is the same throughout.

First, I select three pens of pullets (see selection of breeding stock), and these pens are all three unrelated one to the other. I mate each pen with a cockerel (see selection of cockerels), also unrelated to each other. I trap nest these three pens the first year, and say the records are then as follows:

Pen No. 1
Pullets ring No.........	1	2	3	4
Eggs laid...............	150	170	230	240

Pen No. 2
Pullets ring No.........	5	6	7	8
Eggs laid...............	150	170	230	240

Pen No. 3
Pullets ring No.........	9	10	11	12
Eggs laid...............	150	170	230	240

The eggs laid are hatched out and the progeny toe punched. I take the best cockerel from the best hen (No. 4) and mate him up to the best pullets from hen No. 12, and vice versa, a cockerel from No. 12 to the best pullets from hen No. 4.

I trap nest second year, and say these two pens are numbered:

	Pen No. 4				Pen No. 5			
Ring No..	13	14	15	16	17	18	19	20
Eggs laid..	150	170	200	240	150	170	200	240

Hatch out the eggs, toe punch and mate cockerel, the best one from the best hen, No. 16 to his own sisters from the same hen, No. 16. This makes the progeny related. I trap nest the third year this brother and sister, mating as pen No. 6.

	Pen No. 6			
Ring No................	21	22	23	24
Eggs laid.............	40	60	80	120

I pick out the best pullets from the best-laying hen of this sister and brother mating, No. 24, and mate up with the best cock of the initial pen No. 2 from the best-laying dam, No. 8 hen.

The pullets from this mating are the ones that go into the competitions and make such remarkable egg-laying records.

CHAPTER X

Specially Written for This Work by F. V. L. Turner, Secretary of The North American Egg Laying Competitions

The Sign of a Good Hen—Some Valuable Feed Hints

"Well, the old hen's comb is frozen and she won't lay again for some time."

To those of us who knew the country in our golden youth grandmother's statement carried nothing more than a bit of household news, uninteresting and without value, so far as we were concerned.

Today the most unobservant poultry breeder is brought up with a jerk when the frozen comb digs into the profits of his flock.

It is not the purpose of this chapter to deal with methods preventive of frozen combs and consequent loss in egg production; proper housing is a preventive and the only governing factor in the case.

This chapter is contributed solely as an educational feature, based on the observance of the laying and the non-laying hen's outward characteristic, indicating prolificacy of egg yield.

Our grandmothers knew the ill effects from frozen combs; little did they know, however, of the co-relation between the hen's comb and egg yield, nor the extent to which selection of the layer against the non-layer could be carried toward increased egg yield, through observance of headgear development.

In the photograph of "Baroness II" a remarkable development of comb and wattles is portrayed. Coupled with this development, the characteristics of depth of body, flat back and length of underline all go to make up what has generally been considered in America as the egg type. In addition to the characteristics mentioned, "Baroness II" is possessed of a heavy bone structure, full breast and wide-set legs. Whether or not an egg type exists has not been determined by any laying competition, breeder or experiment station.

What has been determined is that there exists a positive co-relation between comb and wattle development, or headgear, and ovarian activity.

The poor layer pictured laid 60 eggs in nine months: a casual inspection of the headgear will immediately make plain the reason for this article.

Beyond doubt, it is safe to assume that a flock of just matured pullets may be separated into two classes, the poor layers separated from the better. Further determinations as to the laying capacity may be made by close attention to general type, the trap nest and a flock of heavy layers built up through Barron's system of mating.

Reversion to type, according to Mendel's law of heredity, will pull down the total number of good layers each year by the degree to which the laying characteristic is indelibly fixed.

From year to year these poor layers and drones will be reduced through mating according to the system given in this work, and always there will be the greater growth of headgear with the heavier layer.

Closely observing the general make-up of the 1500 layers (500 having been employed each year in the egg laying competition conducted by The North American, a newspaper published in Philadelphia, two years at Storrs Agricultural Experiment Station, Storrs, Conn., and now in its third year at Thorndale, Pa.) this headgear development has ever been a subject of the greatest interest to the writer. That the extent of headgear development is significant is fully established in the case of every individual layer with a high score; equally true it is that a layer with a low egg-laying record is invariably small in headgear growth.

By far the greatest headgear growth is seen in the English type of bird, 35 of which are now engaged in making egg records in the above - mentioned egg-laying competitions.

Having a direct bearing on the subject it is well to state that up to the present moment no basis of calculation is used for animal food values other than the cow's digestibility of ground cereals, whole cereals being calculated by the same process.

This matter of feeds is brought up at this point in order that the minds of poultry raisers may be free of any erroneous impressions as to the value of one ration over another, when each is constructed along the same lines, and because of the utter lack of connection between ovarian activity and headgear perfection with any specific ration. In order to show that the utmost in ovarian activity is not dependent on any specific ration, formulae used in the Missouri Egg Laying Contest and The North American Egg Laying

SINGLE-COMB WHITE LEGHORN
60 Eggs in Nine Months

SINGLE-COMB WHITE LEGHORN
256 Eggs in 49 Weeks

Competition are given herewith. Attention is called to the wide difference between the rations used in the two contests and the ration employed by Barron. The ration used by Barron to rear his layers to the point of maturity differs conspicuously from that used by the American breeder who raised the White Plymouth Rock pullet that ran up the remarkable score of 281 eggs in the first year of the Missouri Contest. Of course, all of the rations given are theoretically adopted to maximum egg production coincident with body maintenance, all of the layers showing, in practice, an actual increase in weight at the end of the competition.

The rations in the Missouri Contest have been rearranged each year; the rations used in The North American Competition have been rigidly adhered to through the three years.

In the Missouri Competition the following was the method of feeding:

Grain Mixture
200 lbs. cracked corn
200 lbs. wheat
100 lbs. oats

Dry Mash
100 lbs. wheat bran
200 lbs. middlings
200 lbs. cornmeal
200 lbs. ground or rolled oats
150 lbs. beef scrap
75 lbs. alfalfa
50 lbs. gluten meal
25 lbs. old process oilmeal
8 lbs. fine table salt
25 lbs. charcoal

One half-pint of grain mixture was fed in the litter in the morning to each pen of five birds, and slightly more at night.

Sprouted oats fed, too.

The dry mash was continually before the birds in the hopper, and at 1 o'clock every day about a handful of the same mash was moistened with either skim or buttermilk and fed in a trough. Each pen also receives daily a handful of sprouted oats, which daily practice will be continued.

In the second Missouri Contest a slight change will be made in the scratch grain mixture; in the summer months one part corn and two parts wheat will be the ration, and in the winter months two parts corn and one of wheat will be fed.

To care for the loss of oats in the grain mixture the ground oats portion of the dry mash will be increased to 300 pounds, the beef scrap will be increased to 250 pounds, the charcoal to 35 pounds.

In the Philadelphia North American Contest the following was the method of feeding:

Dry Mash

Coarse bran..........200 lbs.
Cornmeal............100 lbs.
Gluten feed..........100 lbs.
Ground oats..........100 lbs.
Standard middlings.... 75 lbs.
Fish scrap........... 30 lbs.
Low-grade flour....... 25 lbs.

Scratch Grain

Cracked corn	60 lbs.
Wheat	60 lbs.
Heavy white oats	40 lbs.
Barley	20 lbs.
Kaffir corn	10 lbs.
Buckwheat	10 lbs.
Coarse beef scraps	10 lbs.

For green food in such portions of the year that it cannot be grown in the yard dried beef pulp was used in the following manner: A sufficient quantity, one 16-quart pail for 500 layers, was allowed to absorb all the water possible and fed daily. The coarse beef scrap used in the scratch grain mixture is secured by sifting the ordinary beef scrap; the residue of coarse stuff is then mixed with the scratch grain and placed in the automatic feeders, where the layers work it out in the usual manner, but the automatics are not put into operation until 3 P. M. The dry mash mixture is fed in pans. The mash is before the layers at all times.

As a matter of general information, the high-scoring individual in the first North American event was a Kentucky reared Rhode Island Red, with 256 eggs to her credit. The winning pen of five birds (Barron's White Leghorns) in the second year of The North American affair averaged 238 eggs. Each individual in this pen had the same splendid head development peculiar to "Barron-ess II."

Laying 105 eggs in 119 days, between November 1st, 1913, and February 28th, 1914, the Barred Plymouth Rock pullet shown on page 75 is character-

BARRED PLYMOUTH ROCK

253 Eggs in 47 Weeks

ized by the same headgear development as are the Leghorns or any other breed or variety that lays in excess of the average. Quite a little discussion relative to the laying capacity of the exhibition, or show, bird has been set at rest by laying competitions.

An exhibition specimen of Leghorn is shown on page 71, with a record of 256 eggs in 49 weeks; in this case the headgear shows a development consistent with egg output.

Increasing the egg yield, however, must be accomplished through matings on Barron's lines either with utility or show birds; that is, a foundation for building a laying strain may be just as surely laid with show birds as so-called utility fowls.

Summary of Talk Tom Barron Had With T. E. Friscaberry, Director State Poultry Experimental Station, Mountain Grove, Mo.—Mr. Friscaberry Is in Charge of the Missouri Egg Laying Contest

IS THERE A POSITIVE EGG TYPE?

Everything Indicates That There Is an Egg Type in Poultry, the Same as There Is a Dairy Type in Cattle—"But What Is the Correct Type?"—That is the Question

All signs sometimes fail, and there are exceptions to all rules, but we are of the opinion that there is a definite type in poultry which indicates productiveness the same as a good dairyman finds that certain characteristics and a certain conformation in cattle indicate the productiveness or non-productiveness of certain individuals. Among other things, it is generally agreed that to be a good milker a cow must be healthy and vigorous, she must have a big body or plenty of capacity and she must have a large udder and large milk veins. The more we study the productiveness in poultry the more firmly we are convinced that the same general principles apply to the productive hen: that there is an egg type in poultry and *the* day will come, if it is not now at hand, when we can tell in a general way the good from the bad by certain and definite characteristics. Of course, in the dairy business the "Babcock test" is the final test of the quantity of butter fat, and just so with poultry, the trap nest will perhaps always be the final test.

Mr. Tom Barron, of Catforth, England, recently visited this place, and while here we had him visit seven or eight different yards and point out one or two good and bad hens in each yard. He did so, and after he had left we looked up the records in each case and found that the hens which he claimed were of poor type and poor producers were among the poorest layers, as verified by our records. In one case we had a White Orpington, which had not laid an egg, and she was in a pen with eleven other hens. The hens were driven before him, and at first glance he picked out this hen, which was a blank so far as production was concerned. The hen was in good health and, to the ordinary observer, looked about like any of the others. He pointed out the good and bad in other pens without handling a single hen. The trap-nest records verified his statements.

We believe that any man with reasonable intelligence who studies year after year the question of selection and breeding for egg production, as Mr. Barron has done, will get a certain and definite type firmly fixed in his mind, just as is true in his case as is true with the dairyman. What is this type? That is the question.

We discussed this and other questions with Mr. Barron for several days. While we will not quote his exact words, and we hope not to misquote him, yet we feel safe in saying that he believes the following to be true: An egg-laying strain cannot be produced by inbreeding. In order for a hen to be a good producer she must be in good health and full of vigor. In his own case he has not bred from a male for years which was not bred from hens which laid 200 eggs or over, and he recommends that method. He does not try to overdo the thing, and has not bred for the 300-egg hen and over. His best record was 283 eggs in one year. He breeds for high averages instead of exceptionally high individuals, and this has been true with both pens which he has had in our egg-laying contests at this place.

A good layer usually stands high in front and the back is not level or the rear higher than the front. The best producers usually have large combs, a high tail and a prominent, large, bright eye. Upon handling the birds he finds that most of the best layers have thin, straight pelvic bones; that there is quite a distance between the points of the pelvic bones and the point of the breastbone. This indicates capacity and lots of room for the egg and digestive organs. He likes the wedge shape, rather narrow in front, but wide behind and wide between the legs.

Mr. Barron agrees that the males must be from high laying hens. He also advises looking well to the females and using hens which have shown they are able to make good records. A hen that will not lay well in winter months is discarded by him, for a hen that doesn't lay well in winter will not make a good record, as a rule, and he wishes to breed hens that lay eggs when eggs are highest in price. He uses two males in his breeding pens, alternating them every five days. The males are full brothers, as a rule. He thus gets better fertility, stronger chicks and better hatches, he thinks. Broodiness will ruin the egg record of any hen. We must breed to eliminate that from our flocks as much as possible. There is a blocky, beefy type in every variety, which does not lay. He advises not to trap nest the entire flock, but trap nest at least a few of the most promising ones. Keep accurate records, pedigree the chicks, and results are sure to follow. There must be regularity in feeding. He believes we should feed more moistened mashes, and also believes in some cases that it pays to soak the grain. A hen, in order to make a good record, must produce quite a large number of her eggs in winter months. His experience has been that the first pullets of a brood to begin laying make the best layers, and the first cockerels to crow usually make the best breeders for egg production. Mr. Barron's views coincide largely with our own and with the results which have been obtained at this Experiment Station. We have made a very careful study of these matters for several years, and we had reached the same conclusions about most things, even before talking to Mr. Barron.

13.

www.ingramcontent.com/pod-product-compliance
Lightning Source LLC
Chambersburg PA
CBHW082354220526
45470CB00008B/2742